PAIDEIA MONOGRAPHS

THE SECULARIZATION OF
SCIENCE

HERMAN
DOOYEWEERD

www.paideiapress.ca
www.reformationaldl.org

The Secularization of Science

This English edition, translated by Robert D. Knudsen, is a publication of Paideia Press (3248 Twenty First St., Jordan Station, Ontario, Canada L0R 1S0). Copyright ©2020 by Paideia Press. All rights reserved.

Except for brief quotations in critical publications or reviews, no part of this book may be reproduced in any manner without prior written permission from Paideia Press at the address above.

Unless otherwise indicated, Scripture quotations are from the ESV® Bible (The Holy Bible, English Standard Version®). Copyright © 2001 by Crossway, a publishing ministry of Good News Publishers. Used by permission. All rights reserved.

Paideia Monograph Series Editor: Steven R. Martins

Book Design by: Steven R. Martins

ISBN 978-0-88815-264-0

Printed in the United States of America

Contents

The Relation of Science and Religion	5
The Necessary Struggle	10
The Reformation of Science and Philosophy	15
The Greek Matter-Form Scheme	19
The Scholastic Nature-Grace Scheme	23
The Nature-Freedom Scheme	29
Positivistic Historicism	37
Concluding Remarks	40
About the Author	43

THE SECULARIZATION OF SCIENCE

The Relation of Science and Religion

WHEN REFERENCE IS MADE to secularization, the secularization of science is often forgotten. That is the case because the greater number of Christians who have enjoyed a scientific education lack a clear idea of the connection between scientific thought and religion. The claim is repeatedly made that by its very nature non-theological science must be altogether free of personal belief, because its objectivity would be imperiled the moment it was bound to any presuppositions originating in faith. This idea has been accepted without weighing its consequences and without asking whether it is justified from either a biblical or a critical, scientific point of view.

It is forgotten that the secularization of life would

have been impossible apart from the secularization of science, and that this scientific secularization has taken place under the overwhelming influence of the religious secularization effected by post-Renaissance humanism. We have simply come to regard this situation as a *fait accompli*.

The dangers of our Western secularized science have confronted us anew, however, as we have seen its devastating effect on many Oriental students. Because of their contacts with it, many of them have been torn away from the faith of their ancestors and have become easy prey for nihilism or communism.

Indeed, as it has been said, it is the missionary task of the church to preach the gospel to them! They do not understand, however, the Occidental separation between science and faith. The same secularized science which has dried up their ancestral faith will also wither the seed of the gospel. That is because science, secularized and isolated, has become a satanic power, an idol which dominates all of culture.

It would be a mistake to suppose that this secularization of science is nothing more than the natural result of cultural differentiation. To make this supposition would be to imply, in effect, that religion is only a particular realm of culture. The theory has been advanced, that in primitive society religion was indeed connected with all of life but that in the historical process of cultural differentiation it had to separate itself from all the other social realms. But religion – even apostate religion, that is to say, religion which does not take into account the true religion revealed to us by God in the Holy Scriptures – does not allow itself to be restricted to a special realm of temporal life. Instead, it is the central sphere of human existence, which gives life as a whole its ultimate orientation. Differentiation results in disintegration, if it is not balanced by total integration of life. This total integration can be effected only through religion.

It is paradoxical that this last proposition is upheld by modern sociology, which itself has taken the implications of the secularization of science to their extreme limit. Religion is reduced to a social phenomenon, explained casually by means of a collective consciousness, which is supposed to comprise the solid base for the unity of society.[1] Nietzsche, who had a penetrating insight into the nihilistic consequences of secularized science, said that by means of science man had killed his gods. In his time it was only a prophecy, since science itself was still venerated as a goddess, who would lead humanity along the way of progress, truth, and freedom. At present, however, that proph-

---

1. E.g., in the thought of Emile Durkheim (Tr.)

ecy has come true to a large extent. The faith in the power of science to liberate and to exalt mankind has been undermined and shattered by positivistic historicism and vitalism, both of which have emerged as a result of the radical secularization of modern thought.

Meanwhile, secularized science has never ceased to be the dominating force in Occidental culture. Quite the contrary! Its power has been enhanced to an astonishing degree as it has given rise to unheard-of technological advances. It is an impersonal power which has rationalized all of society. Even if it is no longer venerated as a goddess, it can nevertheless manifest itself as a demon, impressing on man's soul the theoretical image of reality which it has created, an image which cannot be squared with the Christian faith.

It is a vain illusion to suppose that Christian faith has only to do with the world beyond and has nothing to do with science! Secularized science profoundly affects the human heart. From the very moment one accepts it, it accompanies him when he reads the Scriptures and when he says his prayers.

Though the secularization of science was accomplished under the influence of modern, post-Renaissance humanism, it is also necessary to recognize how influential was the central motive of Catholic Scholasticism, that of nature and grace, in preparing the way for

THE SECULARIZATION OF SCIENCE | 9

this later secularization. It is likewise the dominant influence of this anti-biblical and dualistic motive which up to the present day has prevented orthodox-Protestantism from closing its ranks and taking a positive, unequivocal stand against the secularization of science.

What is involved here is not merely a protest against certain clearly un-biblical theses of secularized science; there must be a protest against the entire spirit of secularization as such, of the dogma of the autonomy of science with regard to faith. This spirit and this dogma must be unmasked. What is involved here is no less than an inner reformation of the spirit of science and of its theoretical conception of reality in accordance with the central biblical motive of the Reformation. It is a question of proclaiming that there is a religious antithesis in philosophical and scientific thought, as it was demonstrated in a splendid way by the father of the Calvinistic revival in the Netherlands, Dr. Abraham Kuyper.

We must become aware both of our share of guilt for the secularization of modern science and of our vocation to war against the spirit of apostasy which is revealed in it. That is not to say that we can battle this spirit in our own power. The warfare to which I refer is one of faith, a struggle even with ourselves, in the power of the Holy Spirit, a struggle which finds its dynamic in a life of prayer.

The Necessary Struggle

First of all, we ask why this struggle is necessary from a biblical and from a scientific point of view.

From the biblical point of view we must establish first of all that divine revelation has a central motive, which is the key to knowledge, and that, because of its integral and radical character, this motive altogether excludes any dualistic conception of man and of the world. This motive is that of creation, fall, and redemption in Christ Jesus in the communion of the Holy Spirit. This motive is not at all a doctrine that can be accepted without its working powerfully in our hearts. It is above all a motive force in the very center of our being, the key to the knowledge of God and of the self which can open up to us the revelation of God in the Scriptures and in all the work of his hands. It is a motive which is so central that it is the foundation even for the scientific exegesis of the Scriptures themselves.

This motive is threefold; nevertheless, it is of one piece. It is impossible to understand the truly biblical meaning of sin and redemption without having grasped the true meaning of creation. In revealing himself as the Creator, God reveals himself as the sole origin of all that is. No force can be opposed to him that has any power in its own right. We could not establish any area of terrestrial life as an asylum for our autonomy with reference to the Creator. He has the right to all of our life, to all of our thought, and to all of our action. No sphere of life may be divorced from the service of God. In revealing himself as the Creator, God has at the same time disclosed to man the meaning of his own existence. We are created in the image of God. Taking care to disengage ourselves from all of Greek-inspired speculations of Scholastic theology, that is to say that here God reveals to us the radical unity of our existence.

Just as all of the creation is centered in God as its unified, integral origin, so God has created within man a unitary center, which is the concentration point of his temporal existence with all of its diverse aspects and powers. This is the heart, in the religious sense of the word, the source from which radiate the streams of life, the soul or the spirit of our temporal existence, that is to say, of our bodily existence. For our corporeal existence includes not only the physical aspects and the biological aspects of our being but also the rational aspects and even the temporal function of faith.

Within the heart of man God has concentrated the meaning of all terrestrial reality. That is why the fall of man entails the fall of the entire temporal creation.

That is why, according to the biblical point of view, the world, as it appears in the inorganic, the organic, and the animal kingdoms, cannot be seen as a thing-in-itself independent of man. God has himself revealed to us in his Word that he does not view the creation except with reference to man. It has been marred because of man's sin, and it will be saved by reason of man's redemption.

That is why every philosophy that denies this central place of man in the world is anti-biblical, even when in Scholastic fashion it would hold the macrocosm to be a creation of God. The Thomistic philosophers claim that they unconditionally accept creation in the biblical sense. That is a mistake, however, because they have conceived creation as a truth of the intellect and have interpreted it apart from the key to knowledge.

In connection with the biblical sense of creation, the meaning of the fall also becomes clear. This can be briefly expressed. It is that man, who was created in the image of God, desired to be something in himself, independent of his Creator. Man's self, considered as the individual center of his existence, is, according to the order of creation, destined to reflect the image of God. An image cannot be anything in itself. That is why man's knowledge of himself depends upon his knowledge of God. That is also why human existence, in its religious center, is subject to a law of religious concentration, which has not been abrogated by the fall. All the power of the devil is based on this law of concentration in human existence, because without this law idolatry would be impossible. Sin is a privation, a lie, nothingness; but the power of sin is something positive, which is dependent on the created goodness of reality.

Because man has been created in the image of God, the fall is a radical one, a fall in the religious center, in the very root of human existence, and a fall of the entire world, which has its point of concentration in man. That is also why redemption in Christ Jesus has a radical and integral character. It is the regeneration in Jesus Christ that is at the very heart of our existence. Redemption is in Jesus Christ, who is the new Root of the human race and of the entire earth. In opposition to any dualistic and dialectical conception it is necessary to maintain the radical and integral nature of the creation. That is to say, as Abraham Kuyper had put it, that there is not the least segment of life over which Jesus Christ, the supreme sovereign, cannot claim the exclusive right.

Any theological speculation that attempts to introduce a dialectical tension between the creation and the re-creation in Christ Jesus, between the Word as Creator and the Word as Saviour, is anti-biblical! Neither is there a dualism between common grace and special grace, as if the realm of common grace were separate from the realm of Christ. There is no grace apart from Jesus Christ, the new Root of humanity. The entire domain of common grace is the domain of Jesus Christ. Common grace is nothing more than grace toward mankind as a whole, the humanity which is not yet liberated from its old apostate root, but which is contemplated by God in is new Root, Jesus Christ. It belongs also to the domain of Christ, where the conflict appears between the kingdom of God and the kingdom of darkness. Common grace cannot be interpreted as being the realm of nature, in the Roman Catholic sense, as the autonomous preamble of the realm of grace. On the contrary, it is the sphere of the irreconcilable antithesis between the city of God and the worldly city of the devil.

It is this same religious antithesis which also controls the domain of science and philosophy. Between the central motive of divine revelation and the power of the apostate religious motives conflict is inevitable, since each of them claims to control theoretical thought and the theoretical image of reality. In order to provide a substitute for the secularized conception of reality, it will be necessary for us to discover the theoretical picture of reality that is controlled by the biblical point of view.

The Reformation of Science and Philosophy

To accomplish this inner reformation of science and philosophy, however, it is necessary to obtain a clear idea of the inner point of contact between theoretical thought and the central religious motives which control it at its starting point. From the point of view of the Christian faith, which should subject itself to the central biblical motive in its radical and integral meaning, it is not sufficient merely to reject the autonomy of theoretical reason. The celebrated church father, Augustine, did just that, and he energetically defended the idea that thought cannot find the truth apart from the illumination of divine revelation. It was especially the relationship between philosophy and the Christian religion that he had in mind, and he clearly pointed out the danger of an invasion of Christian thought by Greek philosophy. But such a point of view was never accompanied by a critical investigation of the internal structure of theoretical thought itself. Because he did not grasp clearly the inner point of contact between philosophical thought and religious commitment, Augustine was never able to provide an adequate solution to the problem of a Christian philosophy, properly so-called. He identified the latter question with a totally different one, namely, that of the relationship between philosophy and Christian theology. In denying the autonomy of philosophical thought he also denied the autonomy of philosophy with reference to theology. For him it was impossible to retain the pagan philosophy of the Greeks as an autonomous science. It was necessary to subordinate it to dogmatic theology, considered as the only true Christian philosophy. Philosophy should be accommodated to Christian doctrine. Even though it could be no more than a servant, an *ancilla theologiae*, it could render various services to theology.

We observe, in passing, that this idea of the relationship between philosophy and theology does not at all have a Christian origin. On the contrary, it is the position defended by Aristotle in his *Metaphysics*, when he deals with the relation of metaphysical theology to the other sciences. Aristotle said that theology, the science of the ultimate end and of the supreme good, is the queen of the sciences. The other sciences were not allowed to contradict its axiomatic truths. This Aristotelian thesis was transplanted onto Christian soil and was applied to the relationship of revealed theology to pagan philosophy. But considering his religious starting point, it should go without saying that for Augustine a natural theology in the Aristotelian sense was radically excluded.

THE SECULARIZATION OF SCIENCE | 17

The Augustinian position with reference to Christian science is, therefore, that it is identical with dogmatic theology and that all of the fields of science should be seen from the theological point of view. This position is summarized succinctly in the famous passage in his *Soliloquies: Deum et animum scire volo. Nihil ne plus? Nihil omnino.*[2] It is this position that dominated Scholasticism until the renaissance of Aristotelianism under Albertus Magnus and St. Thomas Aquinas. After this Augustinianism was progressively replaced by the Thomistic conception. Concomitantly a new religious motive made its entrance into Christian thought, one that we have already had occasion to mention, namely, the motive of nature and of grace.

Of course, the terms "nature" and "grace" were already well known. One finds them also in Augustine. But when we speak of a new religious motive, we have in mind a synthesizing motive that tried to reconcile the religious conception of the Greeks concerning nature with the central motive of the Christian religion. That implied that the created world had to be seen under two aspects, one natural and the other super-

---

2. Augustine, *Soliloquies*, I, 7. "God and the soul, that is what I desire to know. Nothing more? Nothing whatever." W.J. Oates, ed., *Basic Writings of St. Augustine* (New York: Random House, 1948), I, 262 (Tr.)

natural. The motive of nature and grace introduces a natural sphere as the autonomous preamble of a supernatural sphere. This supernatural sphere is that of the special revelation of God and of communion with him. In this conception, however, the natural sphere is divorced from the central biblical motive, which we have called the key to all knowledge. The biblical motive is replaced by the religious motive of the Greek conception of nature. Taken in this sense the motive of nature and grace is intrinsically dualistic and dialectical, because it is actually composed of two religious motives which stand in a radical, irreconcilable antithesis to each other. We shall soon examine this situation in more detail.

As we have seen, the central biblical motive of the Christian religion has an integral and radical character, by reason of which it absolutely excludes every dualistic conception of creation. It does not contain, therefore, any vestige of a hidden dialectic. Dualism of whatever sort, any dialectic within the central religious motive controlling the attitudes of one's life and thought, is always born out of an impulse that is partially or totally apostate from this Christian motive.

An apostate motive forces us to seek the absolute within the relative, to isolate an aspect of created reality and to elevate this isolated aspect – which has no meaning except in its universal connection with all the other aspects, and except in its central relation to the divine Origin – to the status of an independent being, which, as a consequence, is deified. What is relative is nothing apart from its correlatives. When an aspect of created reality has been deified, a correlative of this aspect arises with equal force within the religious consciousness; and the absolutization which it engenders sets itself in direct antithesis to that of the deified aspect. Here is the origin of the dialectic within the religious motives which are foreign to the integral and radical position of divine revelation.

The Greek Matter-Form Scheme

We find such a dialectic within the religious motive controlling the Greek view of nature. This is the motive which, after Aristotle, constantly has been called that of matter and form. One of the consequences of the usage of these terms by Scholastic metaphysics, which pretended to be autonomous, was that their religious meaning was completely forgotten. The Greek motive of matter and form has a central, religious character which is impossible to efface in its metaphysical application. It has its origin in an irreconcilable conflict between the older religion of nature and the younger religion of the Olympic gods. In the older religion it is the aspect of organic life which has been deified. The true deity is the eternally flowing vital stream, which cannot be confined to any form whatever, but from which emerge periodically the generations of living beings which have assumed individual form and which are consequently subjected to the fate of death, to unpredictable and pitiless *ananke* (necessity). This religion, which found its typical expression in the cult of Dyonysius, depreciates the principle of form. The divine current of life is unformed, and consequently it is immortal.

Here is the origin of the Greek conception of matter. In the ancient Ionian philosophy *Physis*, nature, is conceived exclusively in this religious sense. *Physis* is the deity itself, the divine Origin of everything that is born with an individual form, the vital stream which flows unceasingly according to the order of time and which survives the death of all finite beings. This is the significance of the mysterious fragment of Anaximander: "Into that from which things take their rise they pass away once more, as is ordained, for they make reparation and satisfaction to one another for their injustice according to the ordering of time."[3]

The meaning of this text can be expressed with the aid of the famous statement of Mephisto in Goethe's

---

3. As quoted by Bertrand Russell, *History of Western Philosophy*, p. 45 (Tr.)

Faust, if one gives it a slight Greek twist:

> *Denn alles was geformt entsteht.*
> *Ist wert das es zu Grunde geht.*

In contrast, the later religion of the Olympic gods arose out of a deification of the cultural aspect of Greek society. It is the religion of form, of measure, and of harmony, which has found its most typical expression in the Delphic Apollo, the law-giver. The Olympic gods left mother earth with its vital flow and its menacing fate of death. They took on ideal personal forms. They became the immortal gods of the city. But they did not have power over the fate which threatened mortal man. Homer says in his *Odyssey*: "For even the immortals cannot aid poor man against cruel destiny."

The Greek motive of divine form stemmed from this cultural religion, and it evoked again as its contrary the motive of matter, the motive of the eternal flux of life and death.

These two antagonistic motives are included within the central dialectical motive of Greek thought. They have continually driven this thought in opposite directions. Every attempt to reconcile them failed, because no one was able to avail himself of a principle which transcended their ultimate antithesis. Since there was no real possibility of a synthesis, the only alternative was to declare the primacy of one motive at the expense of the other. So the ancient philosophy of nature gave the primacy to the principle of matter and depreciated the principle of form. The metaphysics of Plato and Aristotle did the opposite. The god of Aristotle is pure form, and the principle of matter or eternal flux becomes the principle of imperfection, which strives toward a form as the goal of its movement.

This religious antithesis of the motives of form and matter is also expressed in the Greek conception of human nature. Man is composed of rational form and perishable matter. Human nature lacks a radical unity. That is the case because in apostate religion as well as the true religion man's knowledge of himself depends upon his knowledge of God. Since the god of Aristotle is nothing more than a deification of the cultural aspect of form, and since this god is itself confronted with the eternal principle of the alternation of life and death as a power in its own right, man is thought of as being engulfed in the same dualism. That is the reason that the Greek view of nature is incompatible with the biblical view of creation.

Ex nihilo nihil fit: nothing comes out of nothing! This is the essence of Greek wisdom concerning the origin of the world. Precisely for this reason Greek thought can accept the idea of a divine demiurge which gives form to pre-existing matter. The unformed matter itself, however, cannot have its origin in the divine principle of form. The Greek idea of the origin of the world is a dualistic and a dialectical one; and because the Scholastic motive of nature and grace desired to reconcile it with the church doctrine of creation, this motive also found itself enmeshed in a religious dialectic.

The Scholastic Nature-Grace Scheme

It is as it were a general law of such a dialectic, that the religious consciousness first tries to reconcile the ultimate, antithetical elements involved in its ground motive; however, the synthesis disintegrates into the original antithesis as soon as consciousness comes to reflect critically on its starting point. Thomism developed a synthetic conception of the motive of nature and grace. The Ockhamistic and Averrositic nominalism of the fourteenth and fifteenth centuries dissolved this Thomistic synthesis and reduced its terms to a rigid antithesis. In this antithetical view there is no single point of contact between nature and grace.

It is true that William of Ockham gave the primacy again to the motive of grace, which involved depreciating the sphere of the natural until it was conceived to be nothing more than a substratum for the supernatural sphere. Ockham denied that natural reason could attain to metaphysical knowledge and to a natural theology. According to his brand of nominalism, the universals, that is to say, the concepts of genus and species, do not have a real existence apart from the human understanding. They are only signs which stand for the individual things included within their extension; but they do not have any inner connection with them. And because, according to him, science is limited to the knowledge of relations between universals, the criterion of scientific truth is located within the human understanding itself. No matter how much it is depreciated, natural reason is nevertheless completely divorced from divine revelation. It is completely secularized.

Thomistic thought itself attributed a certain autonomy to natural reason; but this autonomy was conceived in a very relative fashion. In fact, according to the synthetic conception of the scholastic motive of nature and grace, natural truth, which are no more than a preamble to supernatural truths, can never contradict the truths of revelation. Scholasticism engages in a continual adaptation of Greek thought to ecclesiastical dogma, a procedure which is completely impossible without a mutual accommodation of the religious motives which dominate these two conceptions of thought.

THE SECULARIZATION OF SCIENCE | 25

As soon as the synthetic conception of the motive of nature and grace was dissolved, and as soon as the two religious motives were again set over against each other in their original antithesis, science could no longer find a place for an accommodation of natural science to ecclesiastical doctrine. The process of the secularization of science had reached its culmination. Christian dogmatic theology, which Augustine and Thomas Aquinas had elevated to the status of a sacred science and which they declared to be the queen of the sciences, was no longer recognized as a science in the true sense of the word. All science was relegated to the sphere of natural reason. The church could indeed condemn the views advanced by secularized science; but here it could not resort to any scientific court of appeal, like it had in theology in its angelic doctors. From this time on even the effectiveness of excommunication depended entirely upon the church's waning political power and on the personal relationship of the scientist to the ecclesiastical authorities.

After the antithetical religious dialectic in the motive of nature and grace had been brought to view, there were two directions in which the science of the Occident could develop. Either Christian thought could return to the central biblical motive and take into account the need for an inner reformation of scientific thought, or the nascent process of the secularization of science could intensify, under the leading of a new religious motive, a product of the complete secularization of the Christian religion. The first possibility offered itself in connection with the great historical movement of the Reformation. The second possibility presented itself in modern humanism, which soon obtained the dominant position in the historical development of our modern culture.

The Reformation could offer no other credentials than the claim to be an inner reformation in a truly biblical sense of the doctrine of the church, of society, indeed of all of life. It was not only a theological and ecclesiastical movement. In calling for a return to the pure spirit of the Holy Scriptures, it summoned forth the driving power of the central biblical motive in its integral and radical meaning, in which it embraces all the spheres of terrestrial life. In the domain of science, the Reformation had, by the grace of God, a great opportunity to effect a basic reform of university instruction in the countries which had aligned themselves with it.

Quite unfortunately the Reformation did not take hold of this opportunity. The magnificent program of Melanchthon for the reform of education was not at all inspired by the biblical spirit. On the contrary, it had a humanistic philological spirit, which was accommodated to Lutheran doctrine and which gave birth to a new scholastic philosophy. The latter, in turn, prepared the way for the humanistic secularization at the time of the Enlightenment. In the Calvinistic universities Theodore Beza restored Aristotelianism as the true philosophy, adapting it to Reformed theology.

This Protestant reform of scientific knowledge cut a miserable figure when it again took up the dualistic maxim: "For faith one must go to Jerusalem; for wisdom one must go to Athens." It was equally discouraging to see in the seventeenth century the celebrated Reformed theologian, Voetius, protesting as a champion of Aristotelianism against the innovations of Descartes. The truly biblical spirit which had inspired John Calvin's *Institutes of the Christian Religion* was conquered by the scholastic spirit of accommodation, which had been imbibed from the anti-biblical motive of nature and grace. It was the driving force of this dialectical motive, the heritage of Roman Catholicism, which stunted the force of the Reformation and which for more than two centuries eliminated the possibility of a serious adversary to the secularization of science.

This secularization was accomplished entirely under the religious influence of modern humanism. It is true that humanism categorically affirmed that the process of secularization was nothing more than a logical outworking of the genius of science itself! That was, however, a very uncritical dogma, which we have unmasked as such in our critical investigation of the inner structure of scientific thought. There has never existed a science that was not founded on presuppositions of a religious nature, nor will one ever exist. This is to say in effect that every science presupposes a certain theoretical view of reality which involves an idea of the mutual relationships which exist between its various aspects, and that this idea, on its own part, is intrinsically dominated by a central religious motive of thought.

Modern humanism, which after the Renaissance more and more dominated the conception of science, itself has a central religious motive, which since Immanuel Kant has been called the motive of nature and freedom. It is impossible for one to understand the ultimate tendencies of the modern secularization of science unless he has obtained a clear view of the religious meaning of this motive. For just as Scholastic thought was deceived because it overlooked the religious nature of the Greek motive of form and matter, so one is also completely deceived about the real nature of the humanistic motive if he thinks that it is no more than the formulation of an exclusively philosophical problem. It is again the influence of the dogma of the autonomy of thought which is responsible for this serious error.

The motive of nature and freedom is a dialectical one. It did not arise, however, out of the collision of two different religions. It arose quite simply from a secularization of the central biblical motive of creation, fall, and redemption.

The Nature-Freedom Scheme

This secularization appeared already at the beginning of humanism in the Italian Renaissance. A purely secular *renascimento* is proclaimed. The biblical conception of regeneration is denatured and becomes the expression of the new humanistic motive of freedom. The latter is no more than a secularization of the biblical theme of freedom in Christ Jesus, the result of redemption. It proclaims the autonomy of man, which is supposed to effect a Copernican revolution in the center of his being, in religion. Human personality is elevated to the position of an ultimate end, a "Selbstzweck," an end in itself. Modern autonomous man wishes to create a God in his own image, whom he can justify in a rational theodicy. Leibniz created a God in the spirit of the humanistic ideal of science, a God who is the great geometer, who can analyze all of reality in an infinitesimal calculus. Here the infinitesimal calculus, which was introduced by Leibniz into mathematics, is deified. Rousseau, who fought passionately against the deification of mathematical science, created a god who corresponded to the sense of freedom of autonomous personality. Immanuel Kant created a god who is the postulate of the practical reason, a god according to the image of an autonomous morality which has proclaimed human personality as its ultimate end.

That there are divergences among the humanistic conceptions of God – which all equally ascribe to him the place of Creator but in a secularized sense – is not simply a happenstance. It reveals a dialectical tension within the central religious motive of freedom. We have said that this humanistic motive arose from a secularization of the biblical theme of freedom in Jesus Christ, regarded as a fruit of redemption. In the Christian religion this motive has a radical sense, because it refers to the root unity of human existence, to the heart, that which transcends the diversity of the various aspects of the temporal order of the world, that in which this entire diversity is concentrated in a spiritual unity which is in the image of God.

As soon as this Christian idea of freedom was secularized, i.e., drawn into the orbit of terrestrial reality with all its variety of aspects and transformed into the humanistic idea of autonomy, it was doomed to become ambiguous. The innate religious tendency which drives one to seek a knowledge of God and of himself thus took an apostate direction. In searching for himself and for his God modern autonomous man is really searching for idols. God as he is revealed in the Holy Scriptures and man as he is created in the image of God are lost to sight.

It is not surprising, therefore, that the religious motive of man's autonomous freedom diverged into two mutually opposing motives, both of which are regarded to be independent and absolute. The motive of autonomous freedom evoked first of all a new ideal of personality with reference to the religious and moral life, an ideal which refuses to be submitted to any practical law which it has not imposed upon itself by its own reason. In the second place, it evoked the motive of the domination of nature by autonomous science and a reconstruction of all reality according to the model of the new natural science founded by Galileo and Newton. That is to say, it evoked the ideal of science.

This new ideal of personality and this new ideal of science which was to dominate the conception of nature both had their origin in the humanistic motive of freedom; but they opposed each other in a dialectical religious tension.

In so far as the theoretical vision of reality was molded by the scientific ideal of the domination of nature, there was no more room for the autonomous freedom of the human personality in the domain of his practical activity. The rationalistic ideal of secularized science developed a strictly deterministic view of reality, deprived of every structure of individuality and construed as a closed, rigid chain of cause and effect.

The new ideal of science secularized the biblical motive of creation. Creative power was attributed to theoretical thought, to which was given the task of methodically demolishing the structures of reality as they are given in the divine order of creation, in order to create them again theoretically according to its own image.

The proud statement of Descartes, repeated by Kant,[4] "Give us material and we shall construct a world for you," and the statement of Thomas Hobbes, that theoretical thought can create just like God himself, are both inspired by the same humanistic motive, the motive of the creative freedom of man concentrated in scientific thought.

Therefore, the very ideal which was evoked by the

---

4. Immanuel Kant, "Allgemeine Naturgeschichte des Himmels", *Immanuel Kant's Werke* (Grossherzog Wilhelm Ernst Ausgabe), II, 267.

THE SECULARIZATION OF SCIENCE | 33

religious motive of creative freedom, the ideal of science in its original naturalistic form, destroyed by its mechanistic theoretical view of the world the very human freedom which had called it forth. On the one hand there was autonomous science, on the other, autonomous action; on the one hand there was the new mathematical and mechanistic ideal of science, on the other, the new ideal of free and autonomous personality. These became mutually antagonistic to each other because of the inner dialectic of the humanistic religious motive. This is what Kant dubbed the conflict between nature and freedom. If one seeks to avoid the dialectical structure of apostate religion, he is faced with the necessity of giving the primacy to one of these two antagonistic motives at the expense of the other.

Just as Greek thought started by giving the primacy to the religious motive of matter – the motive of the unformed, eternal flux of life and of death – so humanistic thought began by giving the primacy to the deterministic ideal of secularized science. It was firmly believed that a secularized and deified science was able to conduct humanity along the road of freedom and of progress.

But with Rousseau there began in the name of freedom a passionate reaction against the ideal of science.

Rousseau depreciated this ideal and gave the religious primacy to the motive of personal freedom embodied in a sentimental religion. Disillusioned he turned away from Occidental culture, which was dominated by science, and proclaimed the regeneration of society by the spirit of freedom.

Kant tried to separate these two antagonistic motives by reserving for each one a domain proper to it. On the one hand, the mechanistic ideal of science was limited to the domain of nature, which had been degraded to the level of a purely phenomenal world, and which was conceived of as a construct of the autonomous understanding of man, the legislator of this world, the origin of natural law. On the other hand, the ideal of autonomous freedom, identified with the idea of pure will, was elevated to the metaphysical status of a norm which transcended the phenomenal world of nature. Within this supersensory kingdom of freedom it was the practical reason which was the autonomous origin of moral law. As was the case with Rousseau, the religious primacy was given to the motive of freedom.

This Kantian idea of the autonomy of the will was conceived in a rationalistic fashion. On the one hand, the true self, the true *autos*, of man was identified with the *nomos*, with the general formulation of the moral law. Within the entire range of his ethics there was no place in Kant for the individuality of the human personality. On the other hand, the humanistic motive of creative freedom could not be content with occupying a purely ideal realm; it could not be content to give over empirical reality, identified with nature, to the rationalistic ideal of science. This motive, just like the motive of the scientific control of nature, had to create a world in its own image.

It is just at these points that Romanticism and post-Kantian idealism wished to eradicate the remnants of rationalism which still clung to the conceptions of freedom and of nature.

A new conception of the ideal of free and autonomous personality was then developed, a conception which no longer sought the true human selfhood, the true *autos*, of man in a general rule, in a moral law, in a *nomos*, but which, on the contrary, considered the true rule of morality to be a simple reflex of the creative individuality of free personality. True morality is, therefore, to follow one's individual disposition and vocation. This new conception of freedom was incompatible with any general law. The "bourgeois morality" and the legalism of Kant were replaced with a "morality of genius." It is impossible to judge a colossus like Napoleon with the same moral rule that applies to an ordinary man!

At the same time there developed a new conception of human society. Under the influence of the mathematical and mechanical ideal of science, society had been dissolved into a congeries of atomistic individuals who were devoid of individuality. It had no room for a conception of community as an individual totality. The new conception of the ideal of free personality, however, which had room only for an individual who was free from every general law, fell into the opposite extreme. It created a universalistic image of society, according to which the individual man is nothing more than a member of a terrestrial individual community, of a totality which completely encloses him and which produces its law and its social order as a reflex of its autonomous individual spirit. According to this irrationalistic view, it is the nations, considered as totalities, which determine the individuality of their members. In such a view there is no more place for the rights of man as such. It is no longer man in general whom one knows; it is only the individual man considered as a member of this nation – Germany, England, France, etc.

In line with this new conception of freedom, there also had to be a remodeling of the conception of nature, which Kant had given over to the rationalistic and mechanistic ideal of science. By means of a dialectical way of thinking, which is not afraid of contradictions, there was the attempt to make a synthesis between the two antagonistic motives which had their source in the religious starting point of humanism. It attempted to discover freedom within nature and natural necessity within freedom.

It is not at all surprising in such a spiritual climate, nourished by the conservative spirit of the Restoration, which dominated the first part of the nineteenth century, that the old ideal of science, suffused as it was by the analytical method of the exact sciences, lost all of its attractiveness.

Positivistic Historicism

A new ideal of science progressively unfolded, one that was oriented to the historical. Just as the mathematical and mechanical model of thought dominated rationalistic philosophy, so this new historical idea of science arose out of the religious humanistic motive of the autonomy of man. But this new historical way of thought was not at all interested in reducing reality to the general formulation of universal laws. On the contrary, it depreciated this rationalistic thought, as one that was incapable of penetrating to the heart of creative individuality. Historical thinking sought its material in unrepeatable individual facts. It wished to interpret them according to their individual character, as belonging to a typical period of development, like the Renaissance, the Enlightenment, the Restoration, etc. And just as the mechanistic and mathematical ideal of science created a mechanistic and rationalistic image of all reality, the new historical ideal of science created a world in its own image. All reality was viewed from the standpoint of history, which was elevated to the position of the absolute. Historical thought created a historical world, at the heart of which there was no more place for other irreducible aspects of life. Nature itself was transformed into an historical nature, a continuous creative evolutionary process. In such a system the cultural history of mankind was considered to be a more advanced state of natural history.

But just as the mechanistic ideal of science was discovered to be antagonistic to the humanistic motive of freedom, so the new historical ideal of science was found to be an even more dangerous adversary to the humanistic ideal of free and autonomous personality. As long as this new historicism was bridled by idealism, which could think of the historical process in no other fashion than as the unfolding in time of the eternal idea of autonomous humanity, historicism could not show its extreme implications.

But post-Kantian idealism, from which historic thought issued, crumbled during the second part of the nineteenth century. Historicism also scrutinized the supposedly eternal ideas of humanism as to their historical aspect, and it reduced them to nothing more than ideological products of the historical process. In emancipating itself from idealism, historicism became positivistic. The biological evolutionism of Darwin and of Marxism transformed historical thought in a naturalistic direction. Both of them possessed an inextinguishable faith in the liberating power of science!

In its turn this religious ideal of secularized science was no longer shielded from the nihilistic implications of extreme historicism. The foundations of the old mechanistic and deterministic ideal of science were broken down at the beginning of the twentieth century, as the result of the discovery of the *quantum* theory of energy.

The hypnosis of Darwinian evolution was followed by a disillusioned awakening, when critical historical research showed that it's *a priori* constructions of the evolution of cultural and social life did not at all agree with the best proven facts. In addition, the two world wars annihilated the faith in the exalting power of science and of autonomous reason.

Faced with all of these facts, positivistic historicism could express itself in its most consistent and extreme form, destroying in its turn the foundations of scientific truth. It nurtured a feeling of decline, which found its philosophical expression in humanistic existentialism and in the famous book of Spengler, *The Decline of the West*.

Concluding Remarks

Thus we have traced to its end the secularization of science in its dialectical development. We have sought to demonstrate that this disastrous process was directed by anti-biblical religious motives, and that neither Roman Catholicism nor Protestantism can absolve itself of its share of responsibility for the development of this secular scientific spirit. They are both responsible for this secularization in so far as they have forgotten the integral and radical nature of the biblical motive and because they have followed the Scholastic motive of nature and grace.

Now we are confronted with the fact that our Western culture has been spiritually uprooted, a state of affairs that is unthinkable apart from the process of the secularization of science.

For the children of the Calvinistic Reformation, there should be no question of wasting time in long scholastic discussions about whether science and philosophy also pertain to the kingdom of Jesus Christ or whether they belong instead to a domain of natural reason. This discussion need not go on, because, as we have shown, there is no natural reason that is independent of the religious driving force which controls the heart of human existence.

For us there are only two ways open, that of Scholastic accommodation, which by reason of its dialectical unfolding results in secularization, or that of the spirit of the Reformation, which requires the inward, radical reformation of scientific thought by the driving power of the biblical motive.

Let us remember the words of our Saviour, "No man can serve two masters." And let us pray to God, that He will send faithful workmen into the harvest field, which is the entire earth, and which therefore includes also the domain of scientific knowledge.

ABOUT THE AUTHOR

Herman Dooyeweerd (1894-1977) was born in Amsterdam to Calvinistic parents whose convictions and way of life were profoundly influenced by Abraham Kuyper, the great Dutch statesman, educator and journalist, and one of the protestant leaders through which the evangelical wing of Dutch reformed protestantism emerged. Dooyeweerd is recorded to have had a prolific career as a researcher in philosophy, during which he wrote various profound literary works such as *The New Critique of Theoretical Thought*, *Roots of Western Culture*, *In the Twilight of Western Thought*, and more. He is, without a doubt, one of the most important philosophers that the Netherlands has ever produced, comparable only perhaps with Baruch de Spinoza.

PAIDEIA MONOGRAPHS

Other Titles (2020–):

The Development of Calvinism in North America
   H. Evan Runner

Point Counter Point
   H. Evan Runner

The Radical Christian Facing Today's Political Malaise
   H. Evan Runner

Christ and Christianity
   Herman Bavinck

The Analogical Concepts
   Herman Dooyeweerd

The Concept of Sovereignty in Modern Jurisprudence and Political Science
   Herman Dooyeweerd

The Criteria of Progressive and Reactionary Tendencies in History
   Herman Dooyeweerd

Looking for more?
Visit www.paideiapress.ca

www.ingramcontent.com/pod-product-compliance
Lightning Source LLC
Chambersburg PA
CBHW050124020526
44112CB00035B/2463